Try It! Even More Math Problems for All

This is not your typical math book.

Try It! Even More Math Problems for All is the third of three collections of offbeat, open-ended problems designed to make even the most math-averse student excited about working through these challenging yet accessible problems.

The **Hints and Solutions** section guides you to probe, suggest, and encourage students to explore even their most unusual insights on the way to solving these 25 new, illustrated problems of varying difficulty. As a result, you will be able to motivate your students to think creatively on their own and to engage in teamwork. And when students solve a problem, you will see and hear their accomplishments.

Perfect for any math classroom, club, after school activity, or coaching session, *Try It!* celebrates not only the destination, but the journey, giving students a chance to think differently, and, above all, have fun!

Can't get enough? Volume 1 in the series, *Try It! Math Problems for All*, and Volume 2, *Try It! More Math Problems for All*, are also available at Routledge.com.

Optional Student Workbook Packs

In addition to this teachers' guide, companion student workbooks are available in packs of ten. The student workbooks feature ample room for student responses and notes, make reviewing and providing feedback on student work easy, provide students with a quick reference to use during discussions, and they save time – there is no need to reproduce student handouts.

Jerry Kaplan is Professor Emeritus at Seton Hall University, USA. He is the author of mathematics books for SRA, Random House, Harcourt Brace, Triumph Learning, and School Specialty.

Other *Try It!* Books Available:

Try It! Math Problems for All
Try It! More Math Problems for All

Try It! Even More Math Problems for All

Jerry Kaplan

Illustrated by Ysemay Dercon

Routledge
Taylor & Francis Group

NEW YORK AND LONDON

Designed cover image: Ysemay Dercon

First published 2024
by Routledge
605 Third Avenue, New York, NY 10158

and by Routledge
4 Park Square, Milton Park, Abingdon, Oxon, OX14 4RN

Routledge is an imprint of the Taylor & Francis Group, an informa business

ISBN: 978-1-032-52413-9 (hbk)
ISBN: 978-1-032-51566-3 (pbk)
ISBN: 978-1-003-40657-0 (ebk)

DOI: 10.4324/9781003406570

Typeset in Bembo
by Deanta Global Publishing Services, Chennai, India

Contents

Preface

To students everywhere,

Hi, I am Jerry and I have 25 new problems. Find one you like and **TRY IT!** That's right. **TRY IT!** We do not ask too much. Just **TRY IT!**

If you cannot figure it out, leave it alone. Then come back to it. Or try another one. And try some more. Speak to a classmate about it. Take your time. Some kids do not get it for a week. That's okay. This is not a race.

After a while, ask for help. A **hint**. Maybe another **hint?**

As a teacher, I spent years challenging students with problems. It started when I was a first-grade teacher. I asked my 19 students to figure out who was the oldest in the class. That took a month to figure out because first all the kids had to find out their exact dates of birth. Then we had to learn new stuff about calendars, years, months, weeks, and days. And finally, we wrote down the names of the oldest and youngest in the class with their birth dates. Later, we made a poster listing the whole class from youngest to oldest. I'll never forget the chatter from the kids about where they were on that list! Very exciting for all. And more: we made lists showing who was born in January, February, March, and so forth. I think I did more teaching in those four weeks than in any other four weeks of my teaching career!

From that experience I realized I wanted to use problem solving in all my classes. I did it later when I taught high school, and continued when I taught in college. Not only did my students have fun and a chance to work together, but students went home and tried our problems on family and friends. And came back to tell stories about all the trouble other folks had solving our problems!

First, take your time. **TRY IT!** and **TRY IT!** again. If you need a **hint**, ask for it. Or, try another problem. Another **hint?** Work on two at the same time. Keep thinking. After a while, you will learn you are better at solving problems than you thought! And that is a lot to learn.

Try It! Hints and Solutions

To teachers everywhere,

This is the third of our series of *Try It!* Just as with the first two, **Try It! Hints and Solutions** is a companion to **Try It! Even More Math Problems for All**. This part of the volume consists of many helpful hints and all the solutions for teachers to use in their classrooms.

The problems. The purposes of these problems are to engage and challenge. We want to give all students a break from the steady routine found in standard math lessons. But even more important, we want students of all backgrounds (and ages) to think and grapple with an array of problems, whether alone or with others.

These problems are different from what you see in schools. They are not connected to the usual lessons or to the yearlong curriculum of any grade. Even more, they are not intended to be inserted at any specific time during the year. They are not aligned to any part of school teachings. Of course, that last might be a stretch since with a bit of effort you can find a place and a time to insert a problem here or there in your program. We suggest that you don't. You will lose spontaneity and the purposes of these problems.

When and where. Use them early or late in the school year. Use them on special days. Use them with one or more students. Small groups make sense. But above all give solvers time to think and brainstorm. Take the problems home? No problem. Remember, there is no clock such as "Your deadline is next Wednesday".

Start anywhere in the set. You will find easy and difficult problems next to each other. Encourage students to write their work in the student booklet. Encourage students to work together and help each other. When a student or a group has a solution, check it quietly. Eventually, you may want students to present their solutions to the class. Some teachers organize a REVEAL DAY for that purpose.

And for you. Have fun presenting problems to your groups – small, large, young, and old. Here is a chance to prod them forward as your audience starts to shout out solutions. By throwing out hints and clues here and there, you encourage more probing. Do not give it away too soon. And above all have fun. Tease kindly. Play.

DOI: 10.4324/9781003406570-1

Background. We used these problems in courses at colleges and in many workshops with teachers and students from 5th to 12th grades. They are used today in schools by many teachers we know. We've presented them to people of all ages – and we all had fun. We continue to get positive feedback.

Be surprised. These problems are for all students, the high and low achievers, the quiet and loud ones. You probably know this: sometimes the quietest and/or the lowest performing students will surprise you if you give them a chance. They surprised us, for sure.

The accent is on **TRY IT!**

A Few Tips

We recognize that it is difficult to translate our personal tastes and styles for other teachers. But there are methods that, when timed correctly, will motivate your students, and encourage them to keep trying when solving problems. In the end, persistence by individuals and groups will solve problems and help to understand the "why" behind the solutions.

Here are a few tips on getting students engaged in solving problems.

- Encourage students to "throw out" ideas, that is, to brainstorm and talk to each other.
- Arrange groups so that students can solve problems with other students.
- "Other students" can be one or as many as four or five 5 students.
- If several groups are working at the same time, keep an eye on every group.
- Keep all students involved.
- Encourage students to explain to other students.
- Do not offer direct help, only tidbits of assistance.
- Keep encouraging students throughout.
- Groups are not in competition; give different problems to different groups.
- Encourage students to write their ideas in the workbook; writing helps when solving problems.
- Do not rush problem-solving sessions. Be patient!

Organization of this Guide

We devote two pages to each problem: A **HINTS** page and a **SOLUTIONS** page. You will find all problems here exactly as you see them in the student workbook.

HINTS: You will find the problem and several hints or prompts on how to get started on the problem.

- Make sure students understand the problem.
- Use these hints only after students understand the problem and had a chance to discuss it within their group.
- Offer hints only after students have made several attempts at solving and seem honestly stuck.
- The longer students struggle with finding a solution, the more they will learn from each other.
- Do not hint too early; that could impede those who are eager to figure things out on their own.
- Suggest that students take a break and come back to the problem later in the day or the next day.

SOLUTIONS: The problem is here again followed by a worked-out solution to the problem.

1 DIVIDING THE CLOCK

Show how to divide a clock face in half so that the sums of the 6 numbers in each half are equal.

What is the sum of each half?

HINTS

What does it mean to divide a clock in half?

You can cut it in half by drawing a line from 12 o'clock to 6 o'clock.

That means there are 6 numbers in one half and 6 numbers in the other half.

Which half do 12 and 6 belong to?

If you choose 12 to belong with the times that follow, you will have these two sets of numbers:

{12, 1, 2, 3, 4, 5} The sum of these 6 numbers is 27.

{6, 7, 8, 9, 10, 11} The sum of these 6 numbers is 51.

1 DIVIDING THE CLOCK

Show how to divide a clock face in half so that the sums of the 6 numbers in each half are equal.

What is the sum of each half?

SOLUTIONS

Cut the clock in half by drawing a line from 9 o'clock to 3 o'clock.

Let 9 belong to the earlier times: {9, 8, 7, 6, 5, 4}

3 now belongs to its earlier times: {3, 2, 1, 12, 11, 10}

Answer: The sum of both sets is 39. Divide the clock from 9 to 3.

2 ODDS OR EVENS

Which is greater – the sum of all the odd numbers or the sum of all the even numbers from 1 to 100? By how much?

HINTS

The odd numbers are 1, 3, 5, 7, ... 97, 99.

The even numbers are 2, 4, 6, 8, ... 96, 98, 100.

Are there more even or odd numbers?

2 ODDS OR EVENS

Which is greater – the sum of all the odd numbers or the sum of all the even numbers from 1 to 100? By how much?

SOLUTIONS

Are there more even than odd numbers?

Match each even and odd number.

1 ----- 2

3 ----- 4

5 ----- 6

.   .

.   .

99 --- 100

There are 50 odd numbers and 50 even numbers.

Answer: In each pairing, the even number is greater by 1. That is true 50 times. So, the sum of the even numbers is greater by 50 than the sum of the odd numbers.

3 SQUARING THE CHECKERBOARD

There are 204 squares in a standard checkerboard, shown below. These squares range from the 64 smallest squares to the largest, which is the checkerboard itself.

How many of these 204 squares contain an equal number of black and white squares?

HINTS

Can you find all the different sizes of squares in the checkerboard? That means the 1 by 1 small squares, the 2 by 2 small squares, the 3 by 3 small squares, etc. up to the 8 by 8 full checkerboard.

This is a sample of a 2 by 2 square.

This sample shows 2 small black and 2 white squares.

3 SQUARING THE CHECKERBOARD

There are 204 squares in a standard checkerboard, shown below. These squares range from the 64 smallest squares to the largest, which is the checkerboard itself.

How many of these 204 squares contain an equal number of black and white squares?

SOLUTIONS

Here is an example of a 3 by 3 square.

Because there are 9 squares in this 3 by 3 square, it is impossible to have an equal number of black and white squares. That leads to this conclusion:

All squares with an odd number of small squares cannot have an equal number of small black and white squares. So, we need to figure out how many squares have an *even* number of small squares.

The table below shows the squares on a checkerboard that have an even number of small squares.

Even number of small squares	How many?
2 by 2	49
4 by 4	25
6 by 6	9
8 by 8	1

Answer: There are 84 squares on a checkerboard that have an equal number of small black and white squares.

4 REVERSING ORDER?

Malcolm says he knows what it means to reverse the digits of a 2-digit whole number. He says 76 is the reverse of 67. Susanna says that he is right. Then she gives Malcolm this problem to solve:

Here are two 2-digit numbers, P and Q:
P has the digits of Q reversed.
Q is 1 less than ½ of P.

Find P and Q.

HINTS

One way to solve this is to write symbols:

$Q = ½ P - 1$

Another way is to think about P and Q is their relative size – 78 and 87 will not work. 62 and 26 will not work.

Search for P:

Are you looking for an odd or even number for P?

Are you looking for a number with the same digits for P, such as 88?

4 REVERSING ORDER?

Malcolm says he knows what it means to reverse the digits of a 2-digit whole number. He says 76 is the reverse of 67. Susanna says that he is right. Then she gives Malcolm this problem to solve:

Here are two 2-digit numbers, P and Q:
P has the digits of Q reversed.
Q is 1 less than ½ of P.

Find P and Q.

SOLUTIONS

Keep trying.

We are looking for P, a 2-digit number without repeating digits.

Try P = 42, Q = 24. Do they work?

No, they do not work because ½ of 42 = 21. 24 is not 1 less than 21.

Keep trying.

Finally, you get P = 52, Q = 25. Do the math: ½ of 52 = 26 and 25 = ½ × 52 − 1

Answer: P = 52, Q = 25

5 FINDING HIDDEN RECTANGLES

How many rectangles are there in this diagram?

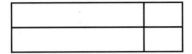

HINTS

Don't give up too soon. Keep looking at this figure.

Take another look. The answer is greater than 4.

5 FINDING HIDDEN RECTANGLES

How many rectangles are there in this diagram?

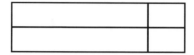

SOLUTIONS

The trick is to "see" the overlapping rectangles.

First, the large rectangle is divided into 4 rectangles.

Along with the large rectangle, that makes 5.

Here's more.

Can you see 2 more rectangles across and 2 more up and down?

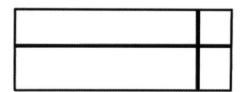

Answer: That makes 9 total rectangles.

6 WATCH YOUR LANGUAGE!

All of Carmen's pets except 2 are dogs.

All her pets except 2 are cats.

All her pets except 2 are turtles.

How many pets does Carmen have?

HINTS

Pick a number such as 5 pets to see how this works.

Go through each line:

With 5 pets, Carmen would have 3 dogs, 3 cats, 3 turtles.

Pick another number and check the result.

6 WATCH YOUR LANGUAGE!

All of Carmen's pets except 2 are dogs.

All her pets except 2 are cats.

All her pets except 2 are turtles.

How many pets does Carmen have?

SOLUTIONS

Keep checking numbers until you get to 3.

Answer: Carmen has 3 pets – 1 dog, 1 cat, and 1 turtle.

7 LOOKING AHEAD?

If there are 5 Mondays, 5 Tuesdays and 5 Wednesdays in January, on what day of the week will February 1st fall?

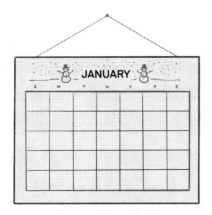 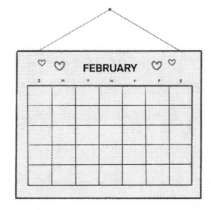

HINTS

Draw a calendar of January with the required number of Mondays, Tuesdays, and Wednesdays.

Check the number of Mondays, Tuesdays, and Wednesdays. Is there another way?

7 LOOKING AHEAD?

If there are 5 Mondays, 5 Tuesdays and 5 Wednesdays in January, on what day of the week will February 1st fall?

SOLUTIONS

Sunday	Monday	Tuesday	Wednesday	Thursday	Friday	Saturday
	1	2	3	4	5	6
7	8	9	10	11	12	13
14	15	16	17	18	19	20
21	22	23	24	25	26	27
28	29	30	31	X		

Answer: February 1st is on a Thursday.

8 REMOVING MATCH STICKS

There are 12 match sticks that make up these 4 small squares.

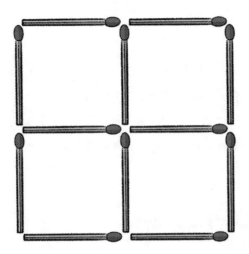

Show your answers to these four instructions with drawings.

  1. **By removing 2 matches, make 2 squares of different sizes.**
  2. **By removing 4 matches, make 2 identical squares.**
  3. **By moving 3 matches, make 3 identical squares.**
  4. **By moving 4 matches, make 3 identical squares.**

HINTS

Use 12 sticks, or 12 pencils, or 12 popsicle sticks, or 12 of anything that can stand for match sticks.

Make a model of the grid and do it.

  1. Remove 2 sticks and make 2 squares of different sizes.
  2. Remove 4 sticks and make 2 squares of identical sizes.
  3. Remove 3 sticks and make 3 squares of identical sizes.
  4. Remove 4 sticks and make 3 squares of identical sizes.

8 REMOVING MATCH STICKS

There are 12 match sticks that make up these 4 small squares.

Show your answers to these four instructions with drawings.

1. By removing 2 matches, make 2 squares of different sizes.
2. By removing 4 matches, make 2 identical squares.
3. By moving 3 matches, make 3 identical squares.
4. By moving 4 matches, make 3 identical squares.

SOLUTIONS

Answers are here in this order:

1. and 2.
3. and 4.

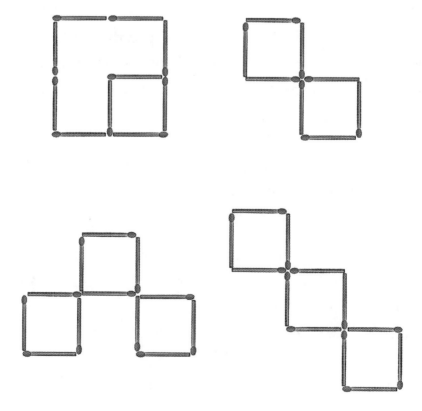

9 GETTING TO 99

Use each digit from 1 to 9 only once to add them together to equal 99. You can use 6 + 7 + 8, or 67 + 8.

$$1 + 2 + 3 + 45$$

HINTS

You can combine digits such as 34 or 61.

Try adding all the digits: $1 + 2 + 3 + 4 + 5 + 6 + 7 + 8 + 9 = 45$.

Which digits combined will help?

Maybe: $1 + 2 + 37 + 4 + 5 + 6 + 8 + 9 = 72$

9 GETTING TO 99

Use each digit from 1 to 9 only once to add them together to equal 99. You can use 6 + 7 + 8, or 67 + 8.

SOLUTIONS

Try a few more:

1 + 2 + 3 + 48 + 5 + 6 + 7 + 9 = 81, not enough.
1 + 2 + 3 + 4 + 58 + 6 + 7 + 9 = 91, closer.

Answer: 2 + 3 + 4 + 5 + 61 + 7 + 8 + 9 = 99.

Is there another way? Yes, 7 more answers with 6 in the tens place.

Here is an example using 67:
1 + 2 + 3 + 4 + 5 + 67 + 8 + 9 = 99

10 NOT A CARD TRICK

There are 3 cards on a desk. These cards have labels, A, B, and C. Each card has a whole number (1, 2, 3, 4, etc.) printed on the other side.

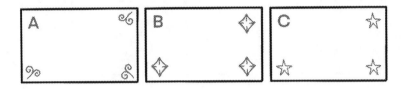

The numbers increase from left to right. That means the card on the left has the smallest number and the card on the right has the largest number.

The sum of the three numbers is 9.

You can turn over only one card. Which card should you turn over to know the numbers of all 3 cards?

HINTS

What sets of three numbers are possible for the cards?
Write these sets of numbers.
If A = 1, then B + C = 8. That means B = 2, C = 6; or B = 3, C = 5: {1,2,6} and {1,3,5}
If A = 2, then B + C = 7. That means B = 3, C = 4: {2,3,4}

10 NOT A CARD TRICK

There are 3 cards on a desk. These cards have labels, A, B, and C. Each card has a whole number (1, 2, 3, 4, etc.) printed on the other side.

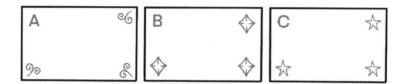

The numbers increase from left to right. That means the card on the left has the smallest number and the card on the right has the largest number.

The sum of the three numbers is 9.

You can turn over only one card. Which card should you turn over to know the numbers of all 3 cards?

SOLUTIONS

There are three sets of increasing numbers that add to 9: {1, 3, 5}, {2, 3, 4}, and {1, 2, 6}.

Here are the numbers of the 3 cards in a table:

A	B	C
1	3	5
2	3	4
1	2	6

If you turn over Card A, you will not know the numbers on the other cards. You might see a 1, and not know the other two numbers.

If you turn over Card B, you will not know the numbers on the other cards. You might see a 3, and not know the other two numbers.

Answer: All three possible numbers for C are different. So, you should turn over Card C. Then you will know the numbers on the other cards.

11 THE WINE BARREL PUZZLE

Ahmed and Malcolm enter the old warehouse. They soon find the 10 barrels they were searching for against a wall. These are old wine barrels, not used for years. They are stacked in the form of a pyramid, 4 at the bottom, 3 on the next level, then 2, and finally 1 on top.

Ahmed and Malcolm notice the numbers on the barrels. They recognize that when you add the numbers along each side of the pyramid, the sums all equal 16.

They find a note that reads: rearrange the barrels so that the sums of the numbers on the pyramid's 2 sides and its base are equal, but also the *smallest* sums possible.

Can you help them solve this riddle?

HINTS

The two sides and the base form a triangle.

When you add the numbers of the base and the other 2 sides, which positions repeat?

How can you get the smallest sums possible? Exchange 7 and 9.

Which 3 numbers do you want at the corners (vertices)? 0, 1, and 2

11 THE WINE BARREL PUZZLE

Ahmed and Malcolm enter the old warehouse. They soon find the 10 barrels they were searching for against a wall. These are old wine barrels, not used for years. They are stacked in the form of a pyramid, 4 at the bottom, 3 on the next level, then 2, and finally 1 on top.

Ahmed and Malcolm notice the numbers on the barrels. They recognize that when you add the numbers along each side of the pyramid, the sums all equal 16.

They find a note that reads: rearrange the barrels so that the sums of the numbers on the pyramid's 2 sides and its base are equal, but also the *smallest* sums possible.

Can you help them solve this riddle?

SOLUTIONS

Place 0, 1, and 2 at the vertices or corners.

Realize that 0, 1, and 2 will each be added twice since they are at the corners.

Exchange 7 and 9.

The total of all numbers to be added is 0 + 0 + 1 + 1 + 2 +2 + 3 + 4 + 5 + 6 + 7 + 8 = 39.

39 ÷ 3 = 13: this means that each sum of the 2 sides and base must be 13.

If the corners are 0 and 1, then the other numbers will add to 12.

If the corners are 0 and 2, then the other numbers will add to 11.

If the corners are 1 and 2, then the other numbers will add to 10.

Answer 1: Base: 0, 7, 5, 1. Side 1: 1, 6, 4, 2. Side 2: 2, 3, 8, 0.

Answer 2: Base: 0, 8, 4, 1. Side 1: 1, 7, 3, 2. Side 2: 2, 5, 6, 0.

12 CALCULATING CALENDAR DAYS

If August 5th falls on a Tuesday, on which day of the week will December 31st fall?

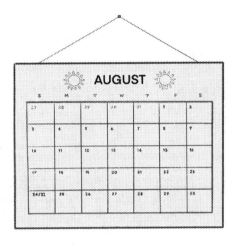
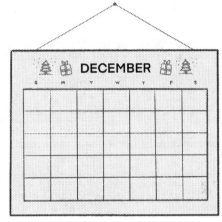

HINTS

Count the days from August 5th to December 31st:

How many days from August 5th to August 31st?

How many days in September?

October?

November?

December?

How many days from August 5th to December 31st?

12 CALCULATING CALENDAR DAYS

If August 5th falls on a Tuesday, on which day of the week will December 31st fall?

SOLUTIONS

How many days from August 5th to December 31st?

MONTH	DAYS
August	26
September	30
October	31
November	30
December	31

26 + 30 + 31 + 30 + 31 = 148 days

148 days = 21 weeks + 1 day

Answer: December 31st is a Wednesday.

13 CROSSING OUT 12 NUMBERS

Here is a grid of 36 numbers. Cross out 12 numbers in the grid to make sure that each row and column has exactly 4 numbers that add to 10.

2	1	2	2	5	4
5	1	1	6	2	3
3	3	5	1	3	1
1	7	6	1	1	2
2	5	1	4	1	3
4	1	2	4	2	3

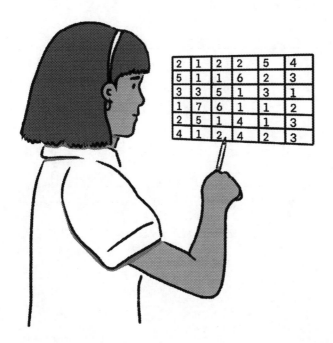

HINTS

Cross out 3 and 4 in the first column.

Cross out 3 and 5 in the second column.

Column 3: the sum of the column is 17.

Remember to leave 4 numbers in each row and column.

13 CROSSING OUT 12 NUMBERS

Here is a grid of 36 numbers. Cross out 12 numbers in the grid to make sure that each row and column has exactly 4 numbers that add to 10.

2	1	2	2	5	4
5	1	1	6	2	3
3	3	5	1	3	1
1	7	6	1	1	2
2	5	1	4	1	3
4	1	2	4	2	3

SOLUTION

2	1	2		5	
5	1	1			3
		5	1	3	1
1	7		1	1	
2			4	1	3
	1	2	4		3

14 FINDING THE NUMBER OF PLAYERS

In one chess league, each competitor played each of the other competitors once and only once. They played a total of 45 games.

How many competitors were there in total?

HINTS

Use a table to get started.

# Players	# Games
2	1
3	3
4	6
5	10
6	15

14 FINDING THE NUMBER OF PLAYERS

In one chess league, each competitor played each of the other competitors once and only once. They played a total of 45 games.

How many competitors were there in total?

SOLUTIONS

Find the pattern in the table on the HINTS page.

The pattern is add 2, add 3, add 4, etc.

Then continue the table:

# Players	# Games
6	15
7	21
8	28
9	36
10	45

Answer: There were 10 players who played 45 games.

15 ADDING CONSECUTIVE NUMBERS

Consecutive means following one after the other in a row. Examples of consecutive letters are m, n, o, p, q. Consecutive numbers are 3, 4, 5, 6 and 75, 76, 77, 78 – numbers that follow one after the other.

If you add 4 + 5 + 6 you get 15. Or 2 + 3 + 4 + 5 = 14. These are examples of adding consecutive numbers.

Find all numbers between 1 and 25 that *cannot* be written as the sum of consecutive numbers.

HINTS

3 can be written as the sum of consecutive numbers: 3 = 1 + 2

So can 18: 18 = 3 + 4 + 5 + 6

Keep testing.

15 ADDING CONSECUTIVE NUMBERS

Consecutive means following one after the other in a row. Examples of consecutive letters are m, n, o, p, q. Consecutive numbers are 3, 4, 5, 6 and 75, 76, 77, 78 – numbers that follow one after the other.

If you add 4 + 5 + 6 you get 15. Or 2 + 3 + 4 + 5 = 14. These are examples of adding consecutive numbers.

Find all numbers between 1 and 25 that *cannot* be written as the sum of consecutive numbers.

SOLUTIONS

2 cannot be written as the sum of consecutive numbers.

How about 4?

5 = 2 + 3

6 = 1 + 2 + 3

7 = 3 + 4

So, we have 2 and 4. Can you guess the next number?

It's 8. 2 + 3 + 4 is too much, 1 + 2 + 3 is too little.

<u>**Answers: Numbers that cannot be written as the sum of consecutive numbers are 2, 4, 8, and 16. They are all based on 2:**</u>

<u>**2**</u>

<u>**2 × 2 = 4**</u>

<u>**2 × 2 × 2 = 8**</u>

<u>**2 × 2 × 2 × 2 = 16.**</u>

16 OPENING AND CLOSING LOCKERS

At the beginning of the school year, 1000 students at Highland Park Middle School stand in line and wait to enter the building. They will enter the first-floor entrance where the student lockers are. The lockers are numbered from 1 to 1000 and are closed.

Students must enter in these special ways:

1. The 1st student opens every locker.
2. The 2nd student closes every even-numbered locker.
3. The 3rd student goes to lockers 3, 6, 9, and so forth, and opens the locker if it is closed; and closes the locker if it is open.
4. The 4th student goes to lockers 4, 8, 12, and so forth, and opens the locker if it is closed; and closes the locker if it is open.

This pattern continues until all 1000 students have passed through the locker ritual. At the end, find the lockers that are open and the lockers that are closed.

HINTS

Think of what happens to a sample of lockers, from 30 to 40, for example.

Locker 30: Student #1 opens, #2 closes, #3 opens, #5 closes, #6 opens, #10 closes, #15 opens, and #30 closes. Locker 30 is closed. Continue testing lockers 31 to 40.

16 OPENING AND CLOSING LOCKERS

At the beginning of the school year, 1000 students at Highland Park Middle School stand in line and wait to enter the building. They will enter the first-floor entrance where the student lockers are. The lockers are numbered from 1 to 1000 and are closed.

Students must enter in these special ways:

1. The 1st student opens every locker.
2. The 2nd student closes every even-numbered locker.
3. The 3rd student goes to lockers 3, 6, 9, and so forth, and opens the locker if it is closed; and closes the locker if it is open.
4. The 4th student goes to lockers 4, 8, 12, and so forth, and opens the locker if it is closed; and closes the locker if it is open.

This pattern continues until all 1000 students have passed through the locker ritual. At the end, find the lockers that are open and the lockers that are closed.

SOLUTIONS

Which numbered students visit locker #35? 1, 5, 7, 35 – an even number of lockers, meaning the locker is closed.

If a locker number such as 35 has an **even** number of factors, it is **closed**.

But if a locker number has an **odd** number of factors, then it is **open**.

The factors of 16 are 1, 2, 4, 8, and 16, an odd number of factors.

All **square numbers** have an odd number of factors.

Answer: The lockers that remain open are ones that are square numbers: 1, 4, 9, 16, 25, 36, 49, and so forth.

17 LOSING TIME?

Wayne's clock was correct at noon, after which it started to lose 17 minutes per hour until six hours ago it stopped completely. It now shows the time as 2:52 p.m. What time is it now?

HINTS

The clock moves forward 43 (60 − 17) minutes every hour.

Think of 43 minutes as its rate: 43 minutes per hour.

The minute hand moved from 12 noon to 2:52. That is 120 + 52 minutes = 172 minutes.

172 minutes translates to 172/43 = 4 hours of actual time.

17 LOSING TIME?

Wayne's clock was correct at noon, after which it started to lose 17 minutes per hour until six hours ago it stopped completely. It now shows the time as 2:52 p.m. What time is it now?

SOLUTIONS

See **HINTS**: The minute hand moved 172 minutes.

172 minutes translates to 172/43 = 4 hours of actual time.

The time shown as 2:52 is 4 hours later.

That means the time was 4 p.m. And that was 6 hours ago.

Answer: The time now is 10 p.m.

<u>18 SPACED OUT?</u>

How would you arrange 4 points so that each point is the same distance from the other 3?

HINTS

See the title of this problem.

18 SPACED OUT?

How would you arrange 4 points so that each point is the same distance from the other 3?

SOLUTIONS

<u>Answer</u>: There is no way to arrange four points (on a plane) that are the same distance from each other. Three points could be arranged in an equilateral triangle, but the only way to place the fourth point would be to move it out of the plane. When we move the fourth point out of the plane to make it equidistant from the three points of an equilateral triangle, we have created the vertices of a regular tetrahedron.

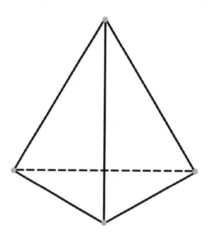

19 OPENING THE LOCK

Possible Codes?

6	8	2
6	1	4
2	0	6
7	3	8
3	8	0

Ahmed has a lock that opens with a 3-digit code. Here are five clues to help you find the correct 3-digit code to open his lock.

Clue 1: For code 682: One digit is right and is in its place.

Clue 2: For code 614: One digit is right and is in the wrong place.

Clue 3: For code 206: Two digits are right, but both are in the wrong place.

Clue 4: For code 738: All digits are wrong.

Clue 5: For code 380: One digit is right and is in the wrong place.

Use the five clues to find the 3-digit code that will open the number lock.

HINTS

Follow each clue to figure out each digit.
You may need to use two clues together.
What do you learn when you combine Clue 2 and Clue 1?
What do you learn when you combine Clue 4 and Clue 1?

19 OPENING THE LOCK

Possible Codes?

6	8	2
6	1	4
2	0	6
7	3	8
3	8	0

Ahmed has a lock that opens with a 3-digit code. Here are five clues to help you find the correct 3-digit code to open his lock.

Clue 1: For code 682: One digit is good and is in the correct place.

Clue 2: For code 614: One digit is good and is in the wrong place.

Clue 3: For code 206: Two digits are good, but both are in the wrong place.

Clue 4: For code 738: All digits are wrong.

Clue 5: For code 380: One digit is good and is in the wrong place.

Use the five clues to find the 3-digit code that will open the number lock.

SOLUTIONS

What did you learn when you combined Clue 2 and Clue 1 (see HINTS)?

Since 6 is in the same place, it will not work. Cross off 6.

What do you learn when you combine Clue 4 and Clue 1?

You learn that 8 in Clue 1 is also out in addition to the 6 of 682, so digit 2 is good and in the correct place.

We have ?-?-2 for the code.

Clue 3 also yields that 0 is good and in the left position.

We have 0-?-2.

Finally, we go back to Clue 2: look at 614. 6 is out; and since the only slot open is in the middle, 4 must be the middle digit.

Answer: The 3-digit code is 042.

20 PERFECT NUMBERS

A perfect number is a number whose factors other than the number itself add up to the number.

Example: The factors of 6 are 1, 2, 3, and 6. The 3 factors other than 6 add up to 6: 1 + 2 + 3 = 6. 6 is a perfect number.

Find another perfect number less than 100.

HINTS

Explain "factors" with examples.

Which numbers can you eliminate?

Prime numbers can be eliminated because their only factors are 1 and itself.

Examples of prime numbers: 3, 5, 7, 11, 13, 17.

Do you think 35 is a perfect number? As you can see, its factors other than itself are 1, 5, and 7.

35 is an example of a number that has few factors. These numbers can be eliminated.

20 PERFECT NUMBERS

A perfect number is a number whose factors other than the number itself add up to the number.

Example: The factors of 6 are 1, 2, 3, and 6. The 3 factors other than 6 add up to 6: 1 + 2 + 3 = 6. 6 is a perfect number.

Find another perfect number less than 100.

SOLUTIONS

You are looking for a number that has factors (not the number itself) that add to that number.

How about 12? Use 1, 2, 3, 4, and 6. Add these and get 16.

Use mental arithmetic to eliminate numbers.

How about 18? Factors are: 1, 2, 3, 6, 9. Sum = 21.

Answer: 28 is a perfect number. Factors are 1, 2, 4, 7, 14.

Note: The next perfect number is 496.

21 REVERSING DIGITS

The sum of the digits of a 2-digit number is 11. If the digits are reversed, the new number is greater than the original by 9. What is the original number?

HINTS

Experiment 1: 74 and 47.

How much greater is 74 than 47? The difference is 27.

Experiment 2: 83 and 38.

How much greater is 83 than 38? The difference is 45.

These two experiments suggest which 2-digit number and its reversal have a difference of 9. Remember, the two digits must add up to 11.

21 REVERSING DIGITS

The sum of the digits of a 2-digit number is 11. If the digits are reversed, the new number is greater than the original by 9. What is the original number?

SOLUTIONS

The digits of the number must add up to 11. 83 and 74 do not work. Neither does 92.

Answer: The original number is 56 (5 + 6 = 11). Reversed it is 65, which is greater than 56 by 9.

Note: Other 2-digit numbers and their reversals differ by 9. For example, 23 and 32; 45 and 54; 87 and 78.

22 CHOOSING FROM TWO PATHS

Here is Salvatore in a park trying to catch up with his friends. He is at a fork in the path and doesn't know which way to go – the left fork or the right fork.

At the fork are twin robots, X1 and X2. One robot always lies. The other always tells the truth. Salvatore can ask the robots only one question.

Which robot should he ask and what question should he ask to make sure he gets on the correct path?

HINTS

If Salvatore asks X1 to point to the correct path, will he know the correct path? Why?

If Salvatore asks X2 to point to the correct path, will he know the correct path? Why?

22 CHOOSING FROM TWO PATHS

Here is Salvatore in a park trying to catch up with his friends. He is at a fork in the path and doesn't know which way to go – the left fork or the right fork.

At the fork are twin robots, X1 and X2. One robot always lies. The other always tells the truth. Salvatore can ask the robots only one question.

Which robot should he ask and what question should he ask to make sure he gets on the correct path?

SOLUTIONS

(1)   Suppose the left fork is the path to the field.

Salvatore points to the left fork while asking this question to X1:
"If I ask X2, is this the correct path, what would it say?"
If X1 tells the truth and X2 lies, then X1 will say that X2 will say no.
If X1 is the liar and X2 tells the truth, then X1 will say that X2 will say no.
So, in both cases the answer "no" means the left path is the correct path.

(2)   Suppose the left path is NOT the correct path.

Salvatore points to the left fork while asking this question to X1:
"If I ask X2, is this the correct path, what would it say?"
If X1 tells the truth and X2 lies, then X1 will say that X2 will say yes.
If X1 is the liar and X2 tells the truth, then X1 will say that X2 will say yes.
In both cases the answer "yes" means the left path is NOT the correct path.

23 FINDING THE AREA OF A PATH

Ezra has a vegetable garden in the shape of a square. He built a path 1 yard wide that goes around the outside of the garden. The area of the path is 40 square yards. What is the area of the garden?

HINTS

Divide the path into 4 parts:

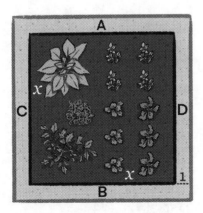

Compute the area of each of the 4 parts and add.

Area of A = ___ [include corners]

Area of B = ___ [include corners]

Area of C = ___ [omit corners]

Area of D = ___ [omit corners]

23 FINDING THE AREA OF A PATH

Ezra has a vegetable garden in the shape of a square. He built a path 1 yard wide that goes around the outside of the garden. The area of the path is 40 square yards. What is the area of the garden?

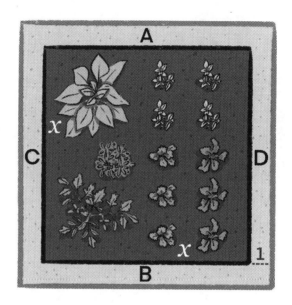

SOLUTIONS

Compute the area of each part:

Area of A = $(x + 2) \times 1 = x + 2$

Area of B = $(x + 2) \times 1 = x + 2$

Area of C = $x \times 1 = x$

Area of D = $x \times 1 = x$

Add the areas of the 4 parts. $4x + 4$

Set this sum equal to the area of the path: $4x + 4 = 40$

Solve for x: $x = 9$. The side of the square equals 9 yards.

Compute the area of the square garden.

Answer: The area of the garden is 81 square yards.

<u>24 CUTTING A SQUARE IN HALF</u>

A square is divided into 16 smaller squares.

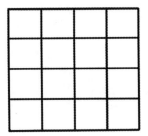

Divide this grid of 16 squares into two halves that are the same size and shape. Show how to do this by making drawings and showing the cuts. The cuts must be along the interior grid lines. There are six ways to do this.

HINTS

Vertical and horizontal cuts (folds) do it. See below.

Recognize that these count as one way since the two resulting shapes are the same.

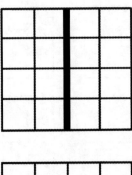

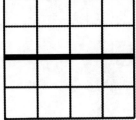

24 CUTTING A SQUARE IN HALF

A square is divided into 16 smaller squares.

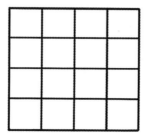

Divide this grid of 16 squares into two halves that are the same size and shape. Show how to do this by making drawings and showing the cuts. The cuts must be along the interior grid lines. There are six ways to do this.

SOLUTIONS

Answers: See the first solution in HINTS. Here are the other five.

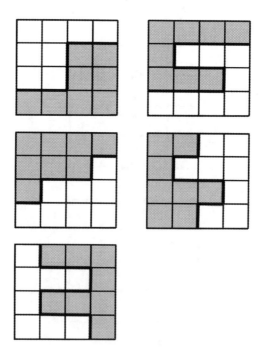

25 THE RIDDLE, OR YOUTH WINS

Felicia is a retired soccer champion. She took her frail husband Todd and their 2 grandchildren Sven, age 6, and Jonah, age 10, on a long hike. They are lost.

It is now dark and raining. They need to cross to the other side of an old broken bridge, but they can cross the bridge only under these circumstances:

- Jonah can cross in 1 minute, Sven can cross in 2 minutes, Todd can cross in 10 minutes, and Felicia can cross in 5 minutes.
- Only 2 can cross at the same time.
- Their only flashlight must be used on all crossings.
- They must all cross in 17 minutes or the bridge will collapse.

How can they all get across safely?

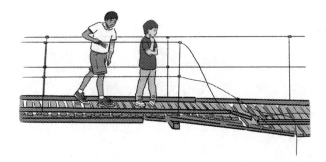

HINTS

How long does it take if Felicia and Todd (wife and husband) cross together? 10 minutes

If Felicia crosses the bridge to return the flashlight, how much time does she take? 5 minutes

Total 15 minutes

This will not work.

See the title of this problem.

25 THE RIDDLE, OR YOUTH WINS

Felicia is a retired soccer champion. She took her frail husband Todd and their 2 grandchildren Sven, age 6, and Jonah, age 10, on a long hike. They are lost.

It is now dark and raining. They need to cross to the other side of an old broken bridge, but they can cross the bridge only under these circumstances:

- Jonah can cross in 1 minute, Sven can cross in 2 minutes, Todd can cross in 10 minutes, and Felicia can cross in 5 minutes.
- Only 2 can cross at the same time.
- Their only flashlight must be used on all crossings.
- They must all cross in 17 minutes or the bridge will collapse.

How can they all get across safely?

SOLUTIONS

Sven and Jonah cross together. 2 minutes

Jonah returns to the starting side. 1 minute

Felicia and Todd cross together. 10 minutes

Total time: 13 minutes

Sven returns to the starting side. 2 minutes

Total time: 15 minutes

Sven and Jonah cross together. 2 minutes

(Note that the flashlight is used on all crossings.)

Total time: 17 minutes

Try It! Even More
Math Problems for All
Photocopiable Problems

What follows are photocopiable versions of all 25 problems. These can be copied and used in group settings or used as handouts where access to a student workbook is unavailable.

DOI: 10.4324/9781003406570-2

1 DIVIDING THE CLOCK

Show how to divide a clock face in half so that the sums of the 6 numbers in each half are equal.

What is the sum of each half?

2 ODDS OR EVENS

Which is greater – the sum of all the odd numbers or the sum of all the even numbers from 1 to 100? By how much?

3 SQUARING THE CHECKERBOARD

There are 204 squares in a standard checkerboard shown below. These squares range from the 64 smallest squares to the largest, which is the checkerboard itself.

How many of these 204 squares contain an equal number of black and white squares?

4 REVERSING ORDER?

Malcolm says he knows what it means to reverse the digits of a 2-digit whole number. He says 76 is the reverse of 67. Susanna says that he is right. Then she gives Malcolm this problem to solve:

Here are two 2-digit numbers, P and Q:

P has the digits of Q reversed.

Q is 1 less than ½ of P.

Find P and Q.

5 FINDING HIDDEN RECTANGLES

How many rectangles are there in this diagram?

6 WATCH YOUR LANGUAGE!

All of Carmen's pets except 2 are dogs.

All her pets except 2 are cats.

All her pets except 2 are turtles.

How many pets does Carmen have?

7 LOOKING AHEAD?

If there are 5 Mondays, 5 Tuesdays and 5 Wednesdays in January, on what day of the week will February 1st fall?

8 REMOVING MATCH STICKS

There are 12 match sticks that make up these 4 small squares.

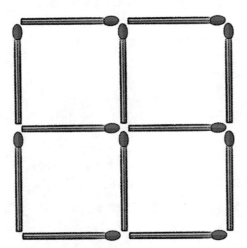

Show your answers to these four instructions with drawings.

1. By removing 2 matches, make 2 squares of different sizes.
2. By removing 4 matches, make 2 identical squares.
3. By moving 3 matches, make 3 identical squares.
4. By moving 4 matches, make 3 identical squares.

9 GETTING TO 99

Use each digit from 1 to 9 only once to add them together to equal 99. You can use 6 + 7 + 8, or 67 + 8.

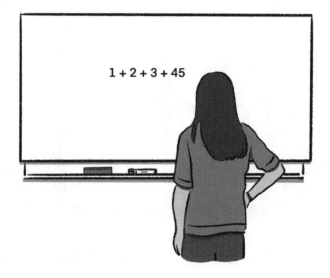

10 NOT A CARD TRICK

There are 3 cards on a desk. These cards have labels, A, B, and C. Each card has a whole number (1, 2, 3, 4, etc.) printed on the other side.

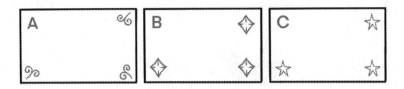

The numbers increase from left to right. That means the card on the left has the smallest number and the card on the right has the largest number.

The sum of the three numbers is 9.

You can turn over only one card. Which card should you turn over to know the numbers of all 3 cards?

11 THE WINE BARREL PUZZLE

Ahmed and Malcolm enter the old warehouse. They soon find the 10 barrels they were searching for against a wall. These are old wine barrels not, used for years. They are stacked in the form of a pyramid, 4 at the bottom, 3 on the next level, then 2, and finally 1 on top.

Ahmed and Malcolm notice the numbers on the barrels. They recognize that when you add the numbers along each side of the pyramid, the sums all equal 16.

They find a note that reads: rearrange the barrels so that the sums of the numbers on the pyramid's 2 sides and its base are equal, but also the *smallest* sums possible.

Can you help them solve this riddle?

12 CALCULATING CALENDAR DAYS

If August 5th falls on a Tuesday, on which day of the week will December 31st fall?

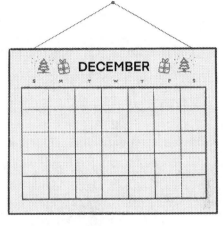

13 CROSSING OUT 12 NUMBERS

Here is a grid of 36 numbers. Cross out 12 numbers in the grid to make sure that each row and column has exactly 4 numbers that add to 10.

2	1	2	2	5	4
5	1	1	6	2	3
3	3	5	1	3	1
1	7	6	1	1	2
2	5	1	4	1	3
4	1	2	4	2	3

14 FINDING THE NUMBER OF PLAYERS

In one chess league, each competitor played each of the other competitors once and only once. They played a total of 45 games.

How many competitors were there in total?

15 ADDING CONSECUTIVE NUMBERS

Consecutive means following one after the other in a row. Examples of consecutive letters are m, n, o, p, q. Consecutive numbers are 3, 4, 5, 6 and 75, 76, 77, 78 -- numbers that follow one after the other.

If you add 4 + 5 + 6 you get 15. Or 2 + 3 + 4 + 5 = 14. These are examples of adding consecutive numbers.

Find all numbers between 1 and 25 that *cannot* be written as the sum of consecutive numbers.

16 OPENING AND CLOSING LOCKERS

At the beginning of the school year, 1000 students at Highland Park Middle School stand in line and wait to enter the building. They will enter the first-floor entrance where the student lockers are. The lockers are numbered from 1 to 1000 and are closed.

Students must enter in these special ways:

1. The 1st student opens every locker.
2. The 2nd student closes every even-numbered locker.
3. The 3rd student goes to lockers 3, 6, 9, and so forth, and opens the locker if it is closed; and closes the locker if it is open.
4. The 4th student goes to lockers 4, 8, 12, and so forth, and opens the locker if it is closed; and closes the locker if it is open.
5. The 5th student goes to lockers 5, 10, 15, and so forth, and opens the locker if it is closed; and closes the locker if it is open.

This pattern continues until all 1000 students have passed through the locker ritual.

At the end, find the lockers that are open and the lockers that are closed.

17 LOSING TIME?

Wayne's clock was correct at noon, after which it started to lose 17 minutes per hour until six hours ago it stopped completely. It now shows the time as 2:52 p.m. What time is it now?

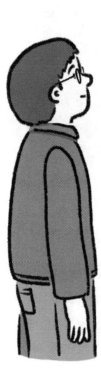

18 SPACED OUT?

How would you arrange 4 points so that each point is the same distance from the other 3?

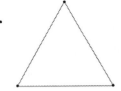

19 OPENING THE LOCK

Possible Codes?

6	8	2
6	1	4
2	0	6
7	3	8
3	8	0

Ahmed has a lock that opens with a 3-digit code. Here are five clues to help you find the correct 3-digit code to open his lock.

Clue 1: For code 682: One digit is good and is in the correct place.

Clue 2: For code 614: One digit is good and is in the wrong place.

Clue 3: For code 206: Two digits are good, but both are in the wrong place.

Clue 4: For code 738: All digits are wrong.

Clue 5: For code 380: One digit is good and is in the wrong place.

Use the five clues to find the 3-digit code that will open the number lock.

20 PERFECT NUMBERS

A perfect number is a number whose factors other than the number itself add up to the number.

Example: The factors of 6 are 1, 2, 3, and 6. The 3 factors other than 6 add up to 6: 1 + 2 + 3 = 6. 6 is a perfect number.

Find another perfect number less than 100.

21 REVERSING DIGITS

The sum of the digits of a 2-digit number is 11. If the digits are reversed, the new number is greater than the original by 9. What is the original number?

22 CHOOSING FROM TWO PATHS

Salvatore is running to play baseball in Eastside Park. This is the first time he has been there. When he comes to a fork in the path, he isn't sure which way to take – the left fork or the right fork.

At the fork are two twin robots, X1 and X2. One of these robots always lies, and the other one always tells the truth. Salvatore can ask only one question to the robots.

Which robot should he ask, and what is the question he should ask to make sure he gets on the right path?

23 FINDING THE AREA OF A PATH

Ezra has a vegetable garden in the shape of a square. He built a path 1 yard wide that goes around the outside of the garden. The area of the path is 40 square yards. What is the area of the garden?

24 CUTTING A SQUARE IN HALF

A square is divided into 16 smaller squares.

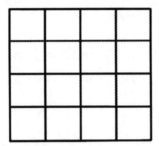

Divide this grid of 16 squares into two halves that are the same size and shape. Show how to do this by making drawings and showing the cuts. The cuts must be along the interior grid lines. There are six ways to do this.

25 THE RIDDLE, OR YOUTH WINS

Felicia is a retired soccer champion. She took her frail husband Todd and their 2 grandchildren Sven, age 6, and Jonah, age 10, on a long hike. They are lost.

It is now dark and raining. They need to cross to the other side of an old broken bridge, but they can cross the bridge only under these circumstances:

- Jonah can cross in 1 minute, Sven can cross in 2 minutes, Todd can cross in 10 minutes, and Felicia can cross in 5 minutes.
- Only 2 can cross at the same time.
- Their only flashlight must be used on all crossings.
- They must all cross in 17 minutes or the bridge will collapse.

How can they all get across safely?

About the Illustrator

Ysemay Dercon is a freelance Illustrator and graduate of the Rhode Island School of Design. She is currently based in Providence, Rhode Island. Her creative pursuits lie in the worlds of publishing, art and science communication, portraiture, and journalistic illustration. She strives to create art that communicates about the world around us in a meaningful and engaging way. You can find her work in books such as *The Little Gardener. Helping Children Connect with the Natural World*, a primer on gardening meant for both children and adults, published by Princeton Architectural Press. She has also worked closely with Osa Conservation, a non-profit conservation and research organization located in the Osa Peninsula of Costa Rica. Her collaboration with Osa Conservation involved creating the illustrations for a series of interpretive panels that have been installed on the trail system of the organization's property. When not creating, she enjoys going for long walks outdoors, reading memoirs, and spending time in cafes with friends.

About the Author

Jerry Kaplan is Professor Emeritus at Seton Hall University. He is the author of mathematics books for SRA, Random House, Harcourt Brace, Triumph Learning, and School Specialty.

Jerry has long advocated for including quality problems as an integral part of good math instruction. This means challenging students with non-routine problems. He did it when he taught first grade. He did it when he taught high school. He did it when he taught college. He urged teachers in his courses and workshops to do the same. This idea became even more important during the pandemic when we heard and read about making up for "lost time." Or "catching up" on lost skills. Jerry thought, "Why not give kids a break and a challenge with problems that are unusual?" So, he found his old class notes and notebooks, and remembered how much fun there was for students and their teachers in solving problems.

Beyond schools, he thinks people at any age can be entertained and challenged with good problems that might take time and help. This volume, the third in a series, is a modest attempt to advance problem-solving for all.